Spinal Interneuron

Plasticity after Spinal Cord Injury

Spinal Interneurons

Plasticity after Spinal Cord Injury

Edited by

Lyandysha Viktorovna Zholudeva
Gladstone Institutes, University of California, San Francisco, CA, United States

Michael Aron Lane
Drexel University, College of Medicine, Department of Neurobiology & Anatomy, and the Marion Murray Spinal Cord Research Center
Philadelphia, PA, United States

Academic Press is an imprint of Elsevier
125 London Wall, London EC2Y 5AS, United Kingdom
525 B Street, Suite 1650, San Diego, CA 92101, United States
50 Hampshire Street, 5th Floor, Cambridge, MA 02139, United States
The Boulevard, Langford Lane, Kidlington, Oxford OX5 1GB, United Kingdom

ISBN: 978-0-12-819260-3

For information on all Academic Press publications visit our website at https://www.elsevier.com/books-and-journals

Publisher: Nikki P. Levy
Acquisitions Editor: Joslyn T. Chaiprasert-Paguio
Editorial Project Manager: Timothy J. Bennett
Production Project Manager: Selvaraj Raviraj
Cover Designer: Miles Hitchen

Typeset by TNQ Technologies

Dedication to Marion Murray (1937–2018): a leader in spinal cord injury research and a pioneer in advancing CNS plasticity

We dedicate this body of work to Dr. Marion Murray, PhD, Professor Emeritus in the Department of Neurobiology and Anatomy at Drexel University, Philadelphia, who passed away on September 9, 2018.

Together with Dr. Michael Goldberger, a close friend and collaborator, Dr. Marion Murray established the Spinal Cord Research Center (SCRC), now the Marion Murray SCRC, at Drexel University. She led the research activities of the Center for over 30 years, served as Principal Investigator of an exceptional NIH Program Project to create one of the most prominent, world-renowned, research centers. She inspired and mentored countless students, fellows and faculty, published more than 150 scientific articles and reviews, and established a tradition of excellence that continues to thrive in the current Department of Neurobiology which we are part of.

An exemplary scientific leader, Marion fearlessly and passionately tackled challenging problems of spinal cord injury to improve human health becoming part of the legacy of women in science over several generations. She had a "Big Picture" vision coupled with skepticism over established dogma in science, and taught us constructive criticism—be it in publishing papers or grant applications—driving the Center to the highest of standards in research and mentoring.

She pioneered research in spinal cord plasticity at a time when neuroscientists were skeptical about axonal sprouting and the neuroplastic potential of the injured spinal cord. This book highlights how much the field now readily accepts the ideas that Marion helped to lay the foundation for. We are extremely grateful for having had the chance to interact with her at Drexel, which contributed to how this book was developed.

We hope that this book will not only serve as a memory to Marion's legacy but would have been a volume she kept on her office bookshelf, frequently read, and shared with her students as she often did with her other books.

We miss you, Marion.

Lyandysha V. Zholudeva and Michael A. Lane
Editors

Contents

Section II
Spinal interneurons – a role in injury and disease

List of contributors

Victoria E. Abraira, Department of Cell Biology and Neuroscience, Rutgers, The State University of New Jersey, Piscataway, NJ, United States; W.M. Keck Center for Collaborative Neuroscience, Rutgers, The State University of New Jersey, Piscataway, NJ, United States

Miriam Aceves, Department of Biology, Texas A&M University, College Station, TX, United States; Texas A&M Institute for Neuroscience, Texas A&M University, College Station, TX, United States

Claudia Angeli, Kentucky Spinal Cord Injury Research Center, University of Louisville, Louisville, KY, United States; Department of Bioengineering, University of Louisville, Louisville, KY, United States; Frazier Rehabilitation Institute, University of Louisville Health, Louisville, KY, United States

Cynthia M. Arokiaraj, Department of Neurobiology, Pittsburgh Center for Pain Research, Center for the Neural Basis of Cognition, University of Pittsburgh School of Medicine, Pittsburgh, PA, United States

Derin Birch, Northwestern University, Feinberg School of Medicine, Chicago, IL, United States

Manon Bohic, Department of Cell Biology and Neuroscience, Rutgers, The State University of New Jersey, Piscataway, NJ, United States; W.M. Keck Center for Collaborative Neuroscience, Rutgers, The State University of New Jersey, Piscataway, NJ, United States

Andrey P. Borisyuk, Neurobiology and Anatomy, Marion Murray Spinal Cord Research Center, Drexel University College of Medicine, Philadelphia, PA, United States; Joint Neuroengineering Initiative, College of Medicine and School of Biomedical Engineering, Science and Health Systems, Drexel University, Philadelphia, PA, United States

Jessica C. Butts, Department of Molecular and Human Genetics, Baylor College of Medicine, Houston, TX, United States

Bo Chen, Department of Neuroscience, Cell Biology, & Anatomy, The University of Texas Medical Branch, Galveston, TX, United States

Steven A. Crone, Cincinnati Children's Hospital Medical Center, Divisions of Pediatric Neurosurgery and Developmental Biology, Cincinnati, OH, United States; University of Cincinnati College of Medicine, Department of Neurosurgery, Cincinnati, OH, United States

Jaclyn H. DeFinis, Marion Murray Spinal Cord Research Center, Department of Neurobiology & Anatomy, Drexel University College of Medicine, Philadelphia, PA, United States

Kimberly J. Dougherty, Marion Murray Spinal Cord Research Center, Department of Neurobiology and Anatomy, Drexel University College of Medicine, Philadelphia, PA, United States

Jennifer N. Dulin, Department of Biology, Texas A&M University, College Station, TX, United States; Texas A&M Institute for Neuroscience, Texas A&M University, College Station, TX, United States

V. Reggie Edgerton, Rancho Research Institute, Los Amigos National Rehabilitation Center, Downy, CA, United States; USC Neurorestoration Center, University of Southern California, Los Angeles, CA, United States; Institut Guttmann. Hospital de Neurorehabilitació, Institut Universitari adscrit a la Universitat Autònoma de Barcelona, Barcelona, Badalona, Spain

Leonardo D. Garcia-Ramirez, Marion Murray Spinal Cord Research Center, Department of Neurobiology and Anatomy, Drexel University College of Medicine, Philadelphia, PA, United States

Yury Gerasimenko, Kentucky Spinal Cord Injury Research Center, University of Louisville, Louisville, KY, United States; Department of Physiology, University of Louisville, Louisville, KY, United States; Pavlov Institute of Physiology of Russian Academy of Science, St. Petersburg, Russia

Simon F. Giszter, Neurobiology and Anatomy, Marion Murray Spinal Cord Research Center, Drexel University College of Medicine, Philadelphia, PA, United States; Joint Neuroengineering Initiative, College of Medicine and School of Biomedical Engineering, Science and Health Systems, Drexel University, Philadelphia, PA, United States

Mark A. Gradwell, Department of Cell Biology and Neuroscience, Rutgers, The State University of New Jersey, Piscataway, NJ, United States; W.M. Keck Center for Collaborative Neuroscience, Rutgers, The State University of New Jersey, Piscataway, NJ, United States

Ngoc T.B. Ha, Marion Murray Spinal Cord Research Center, Department of Neurobiology and Anatomy, Drexel University College of Medicine, Philadelphia, PA, United States

Susan Harkema, Kentucky Spinal Cord Injury Research Center, University of Louisville, Louisville, KY, United States; Frazier Rehabilitation Institute, University of Louisville Health, Louisville, KY, United States; Department of Neurosurgery, University of Louisville, Louisville, KY, United States

Zhigang He, F.M. Kirby Neurobiology Center, Boston Children's Hospital, Boston, MA, United States; Departments of Neurology, Harvard Medical School, Boston, MA, United States

Shaoping Hou, Marion Murray Spinal Cord Research Center, Department of Neurobiology & Anatomy, Drexel University College of Medicine, Philadelphia, PA, United States

Stephanie C. Koch, Department of Neuroscience, Physiology and Pharmacology, Division of Biosciences, University College of London, London, United Kingdom

Michael A. Lane, Drexel University, Department of Neurobiology and Marion Murray Spinal Cord Research Center, Philadelphia, PA, United States

Suh Jin Lee, Department of Neurobiology, Pittsburgh Center for Pain Research, Center for the Neural Basis of Cognition, University of Pittsburgh School of Medicine, Pittsburgh, PA, United States

Ariel J. Levine, Spinal Circuits and Plasticity Unit, National Institute of Neurological Disorders and Stroke, NIH, Bethesda, MD, United States

Erik Z. Li, Marion Murray Spinal Cord Research Center, Department of Neurobiology and Anatomy, Drexel University College of Medicine, Philadelphia, PA, United States

David S.K. Magnuson, Department of Neurological Surgery, Kentucky Spinal Cord Injury Research Center, University of Louisville, Louisville, KY, United States

Amr A. Mahrous, Northwestern University, Feinberg School of Medicine, Chicago, IL, United States

Felicia M. Michael, Spinal Cord and Brain Injury Research Center, University of Kentucky, Lexington, KY, United States; Department of Physiology, University of Kentucky, Lexington, KY, United States

Myung-chul Noh, Department of Neurobiology, Pittsburgh Center for Pain Research, Center for the Neural Basis of Cognition, University of Pittsburgh School of Medicine, Pittsburgh, PA, United States

Bau Pham, Department of Bioengineering, University of California, Los Angeles, CA, United States

Patricia E. Phelps, UCLA, Department of Integrative Biology and Physiology, Los Angeles, CA, United States

Alexander G. Rabchevsky, Spinal Cord and Brain Injury Research Center, University of Kentucky, Lexington, KY, United States; Department of Physiology, University of Kentucky, Lexington, KY, United States

Margo L. Randelman, Drexel University, Department of Neurobiology and Marion Murray Spinal Cord Research Center, Philadelphia, PA, United States

Shelly Sakiyama-Elbert, Department of Biomedical Engineering, University of Texas at Austin, Austin, TX, United States

Alfredo Sandoval, Jr., Department of Neuroscience, Cell Biology, & Anatomy, The University of Texas Medical Branch, Galveston, TX, United States

Rebecca P. Seal, Department of Neurobiology, Pittsburgh Center for Pain Research, Center for the Neural Basis of Cognition, University of Pittsburgh School of Medicine, Pittsburgh, PA, United States

Owen Shelton, Northwestern University, Feinberg School of Medicine, Chicago, IL, United States

Trevor S. Smith, Neurobiology and Anatomy, Marion Murray Spinal Cord Research Center, Drexel University College of Medicine, Philadelphia, PA, United States; Joint Neuroengineering Initiative, College of Medicine and School of Biomedical Engineering, Science and Health Systems, Drexel University, Philadelphia, PA, United States

Alexa Marie Tierno, UCLA, Department of Integrative Biology and Physiology, Los Angeles, CA, United States

Ashley Tucker, Department of Biology, Texas A&M University, College Station, TX, United States; Texas A&M Institute for Neuroscience, Texas A&M University, College Station, TX, United States

Vicki Tysseling, Northwestern University, Feinberg School of Medicine, Chicago, IL, United States

Nicholas White, Department of Biomedical Engineering, University of Texas at Austin, Austin, TX, United States

Lyandysha V. Zholudeva, Gladstone Institutes, University of California San Francisco, San Francisco, CA, United States

Preface

Spinal interneurons have long been recognized as essential components in all spinal networks. They are also now quickly becoming identified as key elements of plasticity with injury or disease, capable of altering their connections and their activity to facilitate—or at times hinder—functional recovery. This book provides a thorough overview of what is presently known about spinal interneurons in the intact and injured spinal cord, presented by leading scientists and clinical professionals.

Preclinical studies of spinal cord neurobiology have recognized their importance in motor, sensory, and autonomic activity since the early work of Santiago Ramon y Cajal and Sir Charles Sherrington. One of the first dedicated texts to the topic of interneurons—"*the Interneuron*," edited by Mary A. B. Brazier—followed a symposium on the topic at UCLA in 1967. This text highlighted the growing interest in spinal interneurons, as scientists recognized their importance. In 1968, G. Adrian Horridge wrote in his book titled *Interneurons*; "What any man or animal perceives or does, and all human conceptions, are restricted by the limitations of interneurons." The present book now reveals the remarkable advances that have been made in this field over the past 50 years, which we hope makes a worthy addition to texts on the subject of "Spinal Interneurons."

As highlighted in the first section of this book, advances in genetic, molecular, electrophysiological, and anatomical assays, with considerations for comparative neurobiology and cross-species differences, continue to reshape our understanding of how spinal interneurons are integrated within and between spinal networks. Ongoing research and characterization of spinal interneurons in developmental neurobiology, computational neuroscience and modeling, systems neuroscience, neurophysiology, and neuroanatomy is expanding the breadth of our knowledge at rapid rates. With these advances also comes the appreciation that the number of identifiable spinal interneurons far surpasses what we had recognized only a decade ago.

As they are essential to function in the intact spinal cord, they are also integral to function after spinal cord injury and disease, as discussed in the second part of this book. Importantly, compromise of the spinal networks with either traumatic injury or degenerative disease rarely results in a total loss of cells or connectivity, and spared neural substrates can contribute to extensive plasticity. Even with the most severe of traumatic spinal cord injuries (e.g., complete spinal transection), spared spinal networks below the level of injury (that lose inputs from the brain) retain the capacity to be activated and functional. The injured spinal cord is now widely recognized as comprising a "new anatomy," rich in interneuronal networks. Although ongoing preclinical and clinical research still aims to better define this new anatomy, there is already widespread recognition that spinal interneurons are crucial components for postinjury function, and as such, represent important therapeutic targets. Their altered neurophysiological function and synaptic connectivity after injury has led to widespread effort to manipulate them with therapeutic intervention, and as described in this book, there has been some considerable success. In fact, spinal interneurons are now often seen as "the gateway to neuroplasticity"—both beneficial and maladaptive—following injury and disease.

The contributions in this book, from some of the world's leaders in spinal interneurons, highlight our current understanding of this highly molecularly, anatomically, and functionally diverse population of cells. Yet it's clear that this is just the tip of the iceberg. We hope that the information here helps to guide increased research effort, harnessing an ever-expanding toolkit for characterizing spinal interneurons and their neuroplastic potential. With a greater understanding of human health and spinal neurobiology, there is hope that novel therapeutic targets will be identified, and that new and improved strategies will be developed for treating spinal cord injury or disease.

Lyandysha Viktorovna Zholudeva, PhD
Michael Aron Lane, PhD

Section I

Spinal interneurons – motor and sensory neuronal networks

Chapter 1

The neuronal cell types of the spinal cord

Stephanie C. Koch[1] and Ariel J. Levine[2]
[1]*Department of Neuroscience, Physiology and Pharmacology, Division of Biosciences, University College of London, London, United Kingdom;* [2]*Spinal Circuits and Plasticity Unit, National Institute of Neurological Disorders and Stroke, NIH, Bethesda, MD, United States*

Introduction

Over the past century, perhaps the most transformative perspective in neuroscience has been to view the function of the nervous system as the coordinated action of diverse populations of neurons and their communication with each other (reviewed in Ref.[1]). Therefore, in our efforts to understand cognition, emotion, somatosensation, and movement, great emphasis has been placed on the need to define the constituent cell types of the brain and spinal cord which could coordinate these complex behaviors.[2,3] Cell "types" are regularly occurring varieties that share a coherent set of cell features such as location, morphology, marker genes, and/or function.

The neuronal cell types of the spinal cord serve as the critical links between the brain and the body. As we navigate through the world, we are constantly using our senses, such as touch, pain, and limb position, to guide how we move and interact with our environment. Through these senses we can protect ourselves from harm and finely correct our movements to adjust to changing conditions. We are able to quickly modify our movements according to environmental needs as a result of specialized networks of cells in the spinal cord, which receive information from the external world, and relay this to motor circuits, as well as to higher command centers.

The spinal cord serves as the first integrative point of primary afferent sensory information with motor networks and spinal circuits are therefore key in the transformation of sensory information and descending signals from the brain into behavior. Within the spinal dorsal horn, sensory input is relayed locally before transmission to higher centers. Here it can be modulated and influenced by both excitatory and inhibitory neurotransmitters released from local interneurons; regulation of these interneurons and the balance between inhibition and excitation can therefore strongly affect the perception and

Spinal Interneurons. https://doi.org/10.1016/B978-0-12-819260-3.00005-6

discrimination of sensory information. Within the ventral horn, spinal cord neurons relay descending cues, coordinate the activity of functionally related groups of muscles, and are sufficient to execute simple behaviors through "central pattern generators" and motor synergies.

Whereas spinal circuit function is carefully managed in the healthy adult system, studies in both animals and humans reveal that the exquisite balance within spinal circuits can be disrupted in pathology, leading to pain and the loss of voluntary movement. For example, removal of spinal inhibition using pharmacological blockade of inhibitory transmitter receptor systems results in excessive neural activity and models of pathological pain states in rats.[4–7] In stark contrast to this, increasing inhibition by means of clinically utilized medicines can dampen down all afferent information from both peripheral tissues and descending from brainstem structures, resulting in anesthesia, analgesia, and paralysis.[8] Direct injury to the spinal cord or loss of brain inputs to the cord following stroke causes paralysis and, paradoxically, spasticity due to exaggerated spinal reflexes.[9] Underscoring the diverse functions of spinal cord cells, spinal cord injury can also cause failure of normal breathing, autonomic regulation, and bowel, bladder, and sexual function.[9]

Advances in the scientific understanding of biological systems have allowed us to take advantage of the versatility of the human organism in order to better treat adult pathologies. Such treatments, however, have thus far proved ineffective and with a large range of side effects, largely due to the lack of knowledge of the underlying specialized circuits in the spinal cord which allow us to sense and adapt to our environment. It remains therefore to not only identify the individual component units of functional behavior, but to understand these units at such a level to target and manipulate them within a functional context. Dissecting the cells and circuits underlying the fundamental interplay between sensation and movement will provide a crucial framework to build upon studying how these processes may in turn go wrong in the case of disease or injury.

Multiple methods have been used to identify and categorize different populations of neurons within the spinal cord. While none of these in isolation have been entirely productive, in combination they have revealed important insights into the cellular basis of coding of sensory stimuli and the execution of behavior.

History of research on spinal cord neurons

Spinal cord neurons and their classification have enjoyed a special place in the history of neuroscience. Santiago Ramon y Cajal formulated the neuron doctrine—that distinct cells are the main anatomical substrate in the nervous system—based partially on his studies of the spinal cord.[10] Around the same time, Charles Sherrington coined the term synapse to describe the specialized connections between neurons and discovered the role of inhibition, both based on his investigations of spinal reflex arcs.[11] Later, his protégé John Eccles performed the first in vivo intracellular recordings of the mammalian central nervous system in cat spinal motor neurons.[12]

These twin traditions of form and function shaped research on spinal cord neuronal cell types for the following decades. Romanes, Rexed, Scheibel, and others provided meticulous analyses of cell type location and morphology, while Eccles, Lundberg, Jankowska, and others probed the electrophysiological properties of cells and traced connectivity within spinal cord circuits.[13–21] By recording from, and subsequently filling (e.g., with a tracer), single identified spinal cord neurons, these researchers also began to link together anatomical and electrophysiological features of spinal cord neurons. These studies have since led to the development and use of a wide array of approaches for characterizing spinal cord neurons, as described below.

Classification systems for spinal cord interneuron cell types

To date, our knowledge of spinal cord neurons has been gained through a range of different techniques, yielding a rich literature and multifaceted knowledge of spinal cord cell types. However, this technical diversity has also fractured our perspectives, with each method revealing cell type specific features and thus providing multiple schemes for classifying spinal cord interneurons (Fig. 1.1).

Anatomy

Anatomy has been a powerful scheme for classifying spinal cord cell types because the location of spinal cord neuron types is intimately linked to their function. Along the rostro-caudal axis, each region of the spinal cord is dedicated to particular outputs such as respiration at mid-cervical levels, hand use at lower cervical levels, and autonomic control at thoracic and sacral levels. Within each spinal cord segment, there is a repeating architecture of histological cell types, organized by Rexed into laminae I through X, each with its own characteristic cellular morphologies.[14] Neurons in the same lamina often share the same descending and sensory synaptic inputs, local connectivity, and extracellular milieu (including neuropeptide volume transmission and signals from glia). However, anatomical organization is an incomplete system for classification, as each lamina contains neurons that vary widely in other cellular features such as neurotransmitter status and firing properties.

Morphology

Morphology has commonly been used as a categorization tool, to some success. Dendritic branching pattern can influence integration capabilities of a

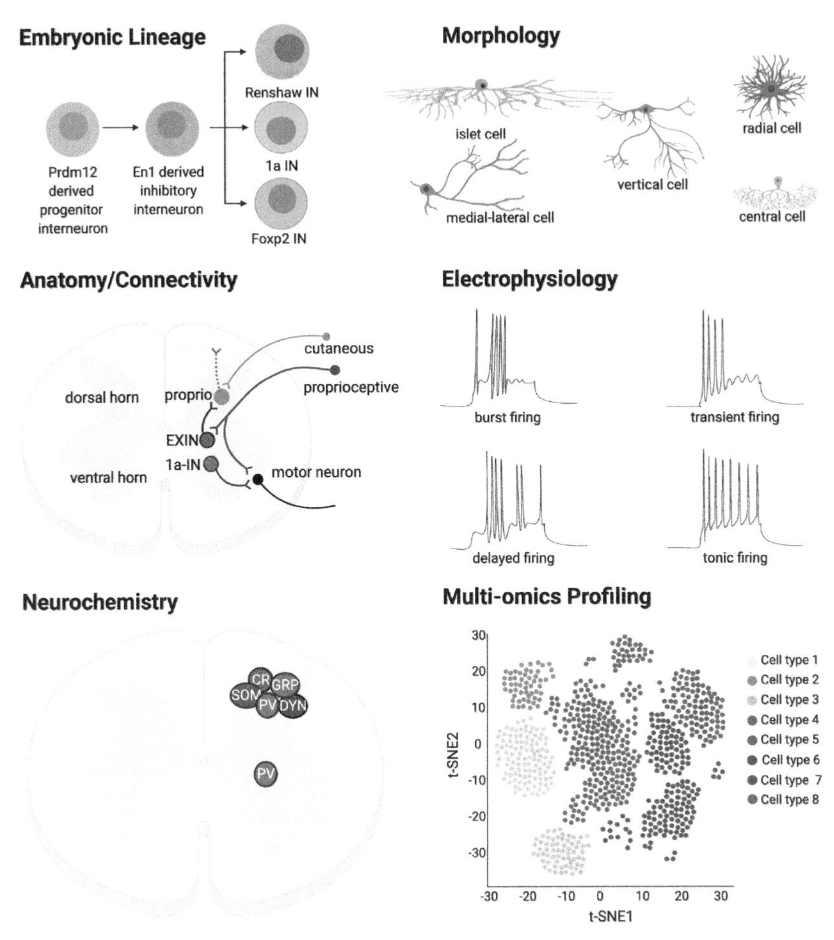

FIGURE 1.1 Summary schematic of the main classifications currently used to catalog spinal interneuronal subpopulations and examples within each categorization. Embryonic lineage tracing, dendritic morphology, anatomy or connectivity, electrophysiological properties, neurochemistry, or multiomics profiling have all been used to subparse spinal interneurons into individual classes, highlighting the diversity of interneuronal populations—although no method has proven to be entirely effective in isolation. See main text for details. *1a-IN*: 1a interneuron; *CR*: calretinin positive interneuron; *DYN*: dynorphin positive interneuron; *En1:* Engrailed 1; *EXIN*: excitatory interneuron; *Foxp2*: Forkhead box protein P2; *GRP*: gastrin releasing peptide positive interneuron; *IN*: interneuron; *Prdm12*: PR domain zinc finger protein 12; *PV*: parvalbumin positive interneuron; *SOM*: somatostatin positive interneuron. *Image created with BioRender.com.*

single interneuron, whether this is confined to a single lamina or across several segments as well as instruct neuronal firing properties.[22,23] Morphological classification has been especially instructive in dorsal horn populations, where interneurons within lamina II of the superficial dorsal horn have been broadly subdivided into islet cell, medial-lateral, radial cell, central cell, and vertical cell morphologies.[22] Within lamina I, neurons have been described to have fusiform, pyramidal, or multipolar morphologies.[24,25] Substantial efforts have been made to match these classifications to functional neurochemical groups and laminar locations, but with limited success.

Connectivity

As primary afferent sensory input enters the spinal dorsal horn, it is subjected to local segmental modulation allowing a tight control on painful, or noxious, information as well as innocuous sensory information. This sensory information entering the spinal cord is organized by sensory modality. Primary afferent terminals exhibit laminar organization depending on their functional class with nociceptive afferents terminating in laminae I–II, low threshold cutaneous afferents terminating in laminae III–V, and proprioceptive afferents terminating in deeper laminae V–VI and within the ventral horn.[26–30] An individual neuron's Rexed laminar position can therefore provide an indication of its role within spinal circuitry. An inhibitory interneuron in the superficial dorsal horn will likely receive predominantly nociceptive afferent input and be involved in the control of noxious sensation, whereas an interneuron in the intermediate dorsal or ventral horn will likely be involved in proprioceptive afferent control or the execution of movement. Within the ventral horn, two classic examples of connectivity-defined neuronal populations are Renshaw cells, which receive collateral input from motor neurons and in turn provide "recurrent inhibition" back to motor neurons, and Ia inhibitory interneurons, which receive primary inputs from Ia sensory afferents and provide output to motor neurons.[19,31]

Importantly however, laminar distribution is not necessarily an accurate gauge of cell activity. This is in part due to the fact that sensory information from peripheral afferents is transmitted both by direct monosynaptic inputs (from primary afferents themselves), and indirect polysynaptic inputs from excitatory interneurons.[32–36] For this reason, electrophysiologists have used the detailed mapping of monosynaptic and polysynaptic input onto individual neurons as a classifier, from the periphery and from neighboring interneurons.[21,37,38]

Electrophysiology

Whereas the anatomical localization of the neuron will determine its primary afferent response profile, intrinsic differences in synaptic chemistry and membrane properties can affect neuronal firing patterns, and so have a strong impact on incoming sensory processing. Electrophysiological properties of neuronal populations are a function of both morphology[22,39] and intrinsic channel expression.[40–43] Firing properties to sustained membrane depolarization range from tonic firing (with little adaptation to the stimulus), to burst firing, and irregular firing. These categorizations have provided some guide to probable function of subpopulations, and a degree of physiological characteristics have been linked to differences in primary afferent input.[42]

Attempts to use electrophysiological responses as a submodality of identified populations have been less successful. There is some evidence for differences in biophysical properties between neurotransmitter phenotypes, with less depolarized action potential thresholds in inhibitory interneurons in the dorsal horn as compared to excitatory vGluT2 positive neurons, leading to less excitable neurons.[44] This, however, could be a function of the mutant mouse and selection of excitatory and inhibitory interneurons reported by the transgene. Similarly firing properties have been found to be somewhat associated with neuronal subtypes, with a higher prevalence of tonic firing in inhibitory interneurons,[22,44,45] although this finding has not been uniformly reported.[27,46] Yet it remains that these examples fail to delineate complexities of types, including anatomical position and neurochemistry.

Neurochemistry

Perhaps the most useful established early categorization was built upon neurotransmitter profile. Within the spinal dorsal horn, excitatory neurotransmission is mediated by glutamate, whereas inhibitory transmission can be glycinergic and/or GABAergic. The influence of inhibitory interneurons on spinal sensory processing can be revealed experimentally by using pharmacological blockade of inhibitory receptors or genetically interfering with inhibitory transmission. Spinal application of $GABA_A$ or glycine receptor antagonists results in increased neuronal firing to cutaneous stimuli,[4,5,47,48] and ablation of GABAergic or glycinergic interneurons results in animals displaying signs of pathological pain and itch states including hyperalgesia, whereby perception of a noxious stimulus is heightened, and touch-evoked allodynia, during which a normally innocuous stimulus is perceived as noxious.[49,50]

Whereas excitatory interneurons and projection neurons form an integrated network involved in the transmission of nociceptive information from primary afferents to higher centers, inhibitory interneurons in conjunction with descending inhibitory influences are involved in the fine modulation and limitation of this flow of information and are activated in parallel to excitatory interneurons. Inhibition through interneurons is achieved both presynaptically by acting directly onto primary afferents, and postsynaptically by inhibiting target neurons. This allows for tight control of incoming signals from the periphery to the spinal cord (feedback inhibition), as well as control on the transmission of information to higher centers (feedforward inhibition), both of which are crucial to dampen excitability of the spinal cord, encourage sensory discrimination, and allow for fine-tuned responses to noxious input. These diverse roles and potential functional outcomes have led to the need for more defined subcategories of excitatory and inhibitory interneuronal populations.

Molecular markers

The use of specific molecular markers greatly advanced knowledge and classification of spinal cord cell types, an approach that accelerated dramatically in the 1990s. Andrew Todd and colleagues used antibodies to neuropeptides and calcium binding proteins to identify diverse cell types in the dorsal horn and to relate features such as morphology, neurotransmitter status, and synaptic inputs.[51–54] With a broad panel of antibodies to transcription factors, Thomas Jessell and colleagues applied this approach to establish one of the best examples of combinatorial molecular codes for neuronal cell types and embryonic lineage.[55–59] Molecular definitions of spinal cord cell types have continued to drive the field forward, with powerful genetic tools to characterize and control specific populations of neurons, and the emerging use of transcriptional profiling to identify spinal cord cell types and their molecular repertoires.

Embryonic lineage

The developmental origin of spinal cord cell types, first characterized with molecular markers, has served as a major classification system for the past 30 years. Overall, a set of transcription factor-defined progenitor domains extend from dI1–dI6 in the dorsal cord to V0–V3 in the ventral cord. Most of these embryonic classes give rise to populations of mature neurons that share neurotransmitter status and general connectivity features, though notable exceptions exist. Lineage is particularly useful in detailing spinal cord neuronal populations derived from the ventral spinal cord but is less effective in explaining diversity in the dorsal spinal cord, in which most neurons derive from two major progenitor domains.

Multiomics profiling

Building on the strength of individual markers for identifying cell types, high content *profiles* of molecular cell type signatures are now being used to define and describe spinal cord cell types. This is really a family of approaches, ranging from combinatorial patterns of antibody labeling that involve dozens of markers to single cell RNA sequencing that reveals the expression of thousands of genes. As this work continues and technologies such as single cell epigenomics and proteomics further expand the toolkit, there will be an explosion of data on the molecular codes of neuronal diversity in the spinal cord.

There are three major strengths of multiomics sequencing strategies as a classification system for spinal cord cell types. First, these methods not only provide lists of distinct spinal cord cell types, they reveal the molecular features that are the raw material of neuronal diversity. Thus, neuropeptides, receptors, and signal transduction components are important to distinguish dorsal horn cell types from each other, while broad patterns of transcription factors are more important in the ventral horn. Second, this data can be linked with cell features that are defined through a range of different perspectives. In this way, we have proposed that transcriptional profiling could form a unifying framework for classifying cell types.[60] For example, tagging cells with genetic markers or circuit tracing labels followed by single cell RNA sequencing has been used to relate embryonic lineage and connectivity with transcriptional profiles.[61–63] In the future, spatial transcriptomics and in situ sequencing will allow us to precisely link cell location and anatomy with gene expression patterns. Third, multiomics profiling has the potential to be a complete classification system because all cell types can be analyzed simultaneously with the same approach and can then be compared with each other. A "cell atlas" of the spinal cord provides a census of cell types and a "map" of how they relate to each other in multidimensional gene expression space.[64] Recent work has provided atlases of the neonatal, adolescent, and adult mouse spinal cord.[65–67] Though we do not yet have a finalized atlas of spinal cord cell types, a major principle that is evident from these studies is the importance of location for spinal cord cell type organization. Neuronal populations from the dorsal horn separate well from each other, forming discrete clusters that represent different cell types, while neuronal populations from the ventral horn are more similar to each other, perhaps existing as overlapping continua of related cell types.

Perspective

The field of spinal cord biology needs a consensus about how to define cell types before we can establish a reference set of cellular components. Our goal may be a single unified system of classification in which each neuronal cell type has a coherent identity with a particular location, embryonic lineage, connectivity, electrophysiological properties, function, and molecular markers. However, it is possible that spinal cord biology is not organized this rigidly. We must be aware of cell types and cellular features that represent exceptions to the major trends. We must also recognize that biological systems are not frozen in time. The relationships between cell types, their defining features, and function may vary throughout developmental time, between different behavioral tasks, and in disease and injury.

Below, we present a summary of spinal cord cell types. As anatomy plays a dominant role in the functional organization of the spinal cord and the classification of spinal neurons, this summary is divided into separate sections for the dorsal horn and the ventral horn of the spinal cord.

The dorsal horn neurons of the spinal cord

Functional subpopulations of spinal dorsal horn interneurons (INs) have begun to be dissected in recent years through the advent of genetic tools and

WG52 and WG53

WC63 and WC62